Yum - Yum!

MICK MANNING AND BRITA GRANSTRÖM

FRANKLIN WATTS

A Division of Grolier Publishing

NEW YORK • LONDON • HONG KONG • SYDNEY

DANBURY, CONNECTICUT

One night,
out of the soil a tiny shoot
pushes up its head...

Who'd eat a tiny shoot?

The shoot is the first part of a plant to grow above the surface.

Lots of animals would — but caterpillar gets there first. Yum-yum!

Caterpillar chews up tiny shoot and then crawls across the meadow.
Who'd eat a caterpillar?

Different caterpillars eat different sorts of plants.

Lots of animals would —
but cricket gets there first.
Yum-yum!

Cricket chomps caterpillar,
then hops off through
the grass.
Who'd eat a cricket?

Crickets call in the evening by rubbing their wings together.

Lots of animals would —
but spider gets there first.
Yum-yum!

Spider gobbles up cricket and
then dangles by a thread.
Who'd eat a spider?

Some spiders build webs to trap their prey; others hunt on foot!

Lots of animals would — but
lizard gets there first. Yum-yum!

Lizard eats up spider, then
sunbathes on a stone.
Who'd eat a lizard?

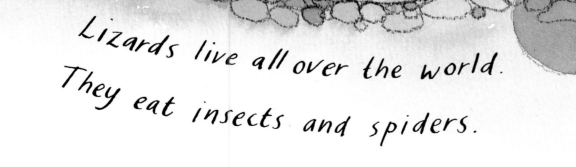

Lizards live all over the world.
They eat insects and spiders.

Lots of animals would – but little owl gets there first. Yum-yum!

Owl swallows lizard, then goes
to drink water from the lake.
Who'd eat a little owl?

Little owls often hunt in the daytime;
they are very small, about the
size of a can of cola!

13

Lots of animals would —
but pike gets there first.
Yum-yum!

Pike snaps up little owl, then lazes
around on the sunny surface.
Who'd eat a big pike?

Pike will eat other fish, birds, and rats!

Lots of animals would — but osprey
gets there first. Yum-yum!

Osprey munches pike, then dozes
off on a dead branch.
Who'd eat an osprey?

Ospreys and fish eagles dive into
the water to catch their prey.

Lots of animals would — but
fox gets there first.
Yum-yum!

Fox gulps down osprey, but he gets a bone stuck in his throat! Fox coughs and coughs and coughs, then fox dies.
Who'd eat a fox?

Foxes will catch food when they can – but they will also scavenge by roads and in dustbins.

Lots of animals would —
but bluebottle flies and
beetles get there first.
Yum-yum!

Bluebottle flies and beetles lay eggs.
Maggots hatch and they eat fox all
up, skin and all, until all that's left
are foxy specks that sink into the soil!
Who'd eat foxy specks in the soil?

Maggots clear up dead animals and rotten food by eating everything up!

Lots of animals would — tiny animals that live in the soil. Yum-yum!

Tiny animals nibble the foxy specks,
putting good things in the soil.
Who'd eat the good things in the soil?

These small animals are too tiny to see without a microscope!

Lots of plants would, but tiny seed gets there first with its root. Yum-yum!

Tiny seed eats the good things in the soil.
Then one day out of the soil a
tiny shoot pushes up its head.
Who'd eat a tiny shoot ?

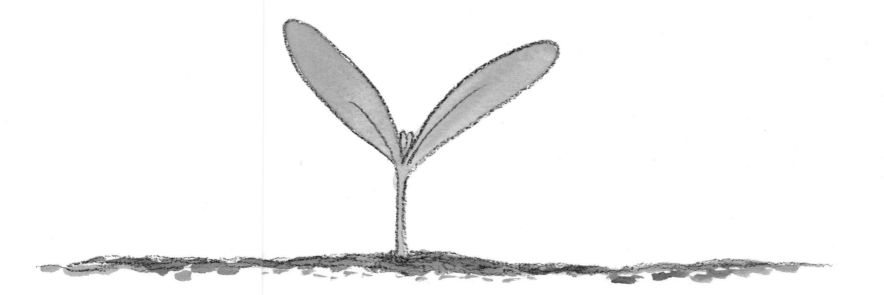

It's a race for seeds to grow the fastest and
make use of the goodness in the soil.

Lots of animals
would — but this
time human being gets
there first!

The green shoots grow into wheat.

We make bread from wheat.

28

29

Helpful words

Bluebottle is a type of fly that lays its eggs on dead animals (pages 20, 21, 29).

Food chain is how a plant and animals are linked together through feeding.

Foxy specks are tiny bits of bone and fur that sink into the ground (page 21).

Maggots are the young of bluebottle flies (pages 20-21).

Magnify is to make something look bigger.

Microscope is a machine for magnifying and looking at tiny living things (page 23).

Predator is an animal that hunts other animals.

Prey is an animal who is hunted for food by another animal (page 9).

Shoot is the first leafy stage of a baby plant (page 3, 28, 29).

Soil, also called earth, is made up of tiny specks of stones, plants, leaves, and dead animals (page 22, 23).

Tiny living things this includes tiny animals such as springtails and mites, and microbes which are sometimes called microorganisms. They are so small you need a powerful microscope to see them (page 22, 29).

For Sue, Steve, Tom,
Emily, and Alex

© 1997 Franklin Watts
First American Edition 1997 by
Franklin Watts, A division of Grolier Publishing
90 Sherman Turnpike, Danbury, CT 06816
Text and illustrations © 1997 Mick Manning
and Brita Granström
Series editor: Paula Borton
Art director: Robert Walster
Consultant: Peter Riley
Printed in Singapore
Library of Congress Cataloging-in-Publication Data
Manning M.
     Yum-Yum! / Mick Manning and Brita Granström
        p. cm. - - (Wonderwise)
     Summary : A simple description of the food chain begins with a
plant which is eaten by a caterpillar which is eaten by a cricket
which is eaten by a spider which is eaten by a lizard ...
     ISBN 0-531-14484-4 (hardcover). -- ISBN 0-531-15322-3 (pbk.)
        1. Food chains (Ecology) -- Juvenile literature. [1.Food chains
(Ecology) 2. Ecology.] I. Granström, Brita. II. Title.
QH541. 14. M345   1997
577'. 16--dc21   97-10726
                                               CIP
                                               AC